ELEMENTS GO TO School TOO!?

This book belongs to: _____

Library of Congress Cataloging-in-Publication Data
        Turpin-Plowden, Jada
Certain trademarks are used under license

Printed in the United States of America
Book design by Jada Turpin-Plowden & Jenesis Scott
Character design Josue' Augustin Advincola

Sage Mastery House LLC

This book is dedicated to the people throughout my life who've dreamed for me while I was awake.

So thank you for nurturing my aptitude when I needed it the most; Ms. Cepeda, Cynthia Loran, Jenesis Scott, Brandon Hammond, My Family and Mom.

Hello! My name is Ms. Emily. Today I'll be showing you around Element Charter.

Element charter is a school specifically designed for chemical elements.

Let's start the tour!

Element
Charter

23
V
...dium
...n Metal

72
Hf
Hafniu...
Transition ...

86
Rn
Ra...n
...le Gas

First, it's important to know exactly what **CHEMICAL ELEMENTS** are. Elements are any substance that cannot be made into smaller substances by a normal chemical process. They are the reason all matter is composed.

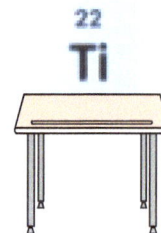

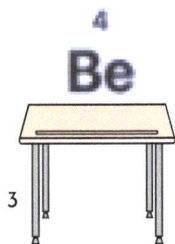

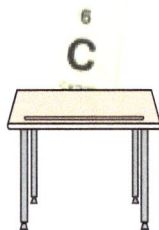

11
Na

14
Si

20
Ca

59
Pr

22
Ti

4
Be

6
C

87
Fr

13
Al

1
H

3

Scientists use Chemical Elements to learn more about the world we live in.

Come meet my class,
I teach period 2.
All the elements in
my class have 2 outer
energy shells.

Does anyone want to volunteer to explain this period?

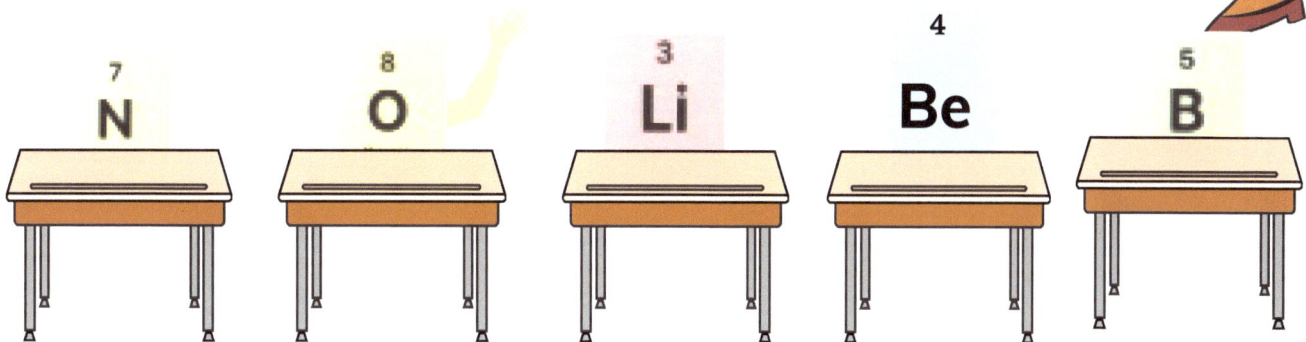

N 7

O 8

Li 3

Be 4

B 5

Hi my name is Oxygen!
I can be found in the air
around you.

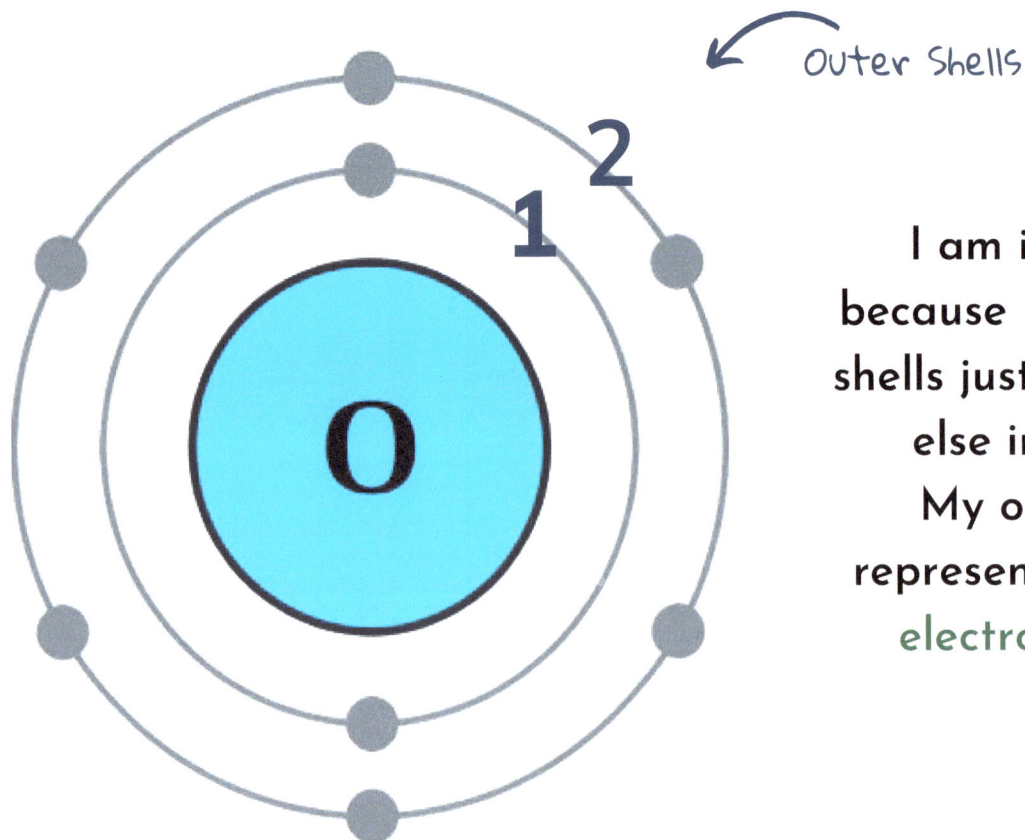

Outer Shells

2
1

O

I am in period 2
because I have 2 outer
shells just like everyone
else in my class!
My outer shells
represent the valence
electrons I carry.

FUN FACT

A Period is a horizontal row of the periodic table.

There are seven periods in the periodic table, with each one beginning on the left side. A new period begins when a new principal energy level begins filling with electrons. Period 1 has only two elements (hydrogen and helium), while periods 2 and 3 have 8 elements. Periods 4 and 5 have 18 elements.

Periods 6 and 7 have 32 elements. The two bottom rows that are separated on the periodic table belong to periods 6 and 7.

Let's go meet some of the other classes!

Hi, I'm Ms. Julia! I teach period 4.

K

All of the elements in my class have 4 outer shells.

24
Cr

20
Ca

21
Sc

19
K

23
V

Let's count the 4 outer shells!

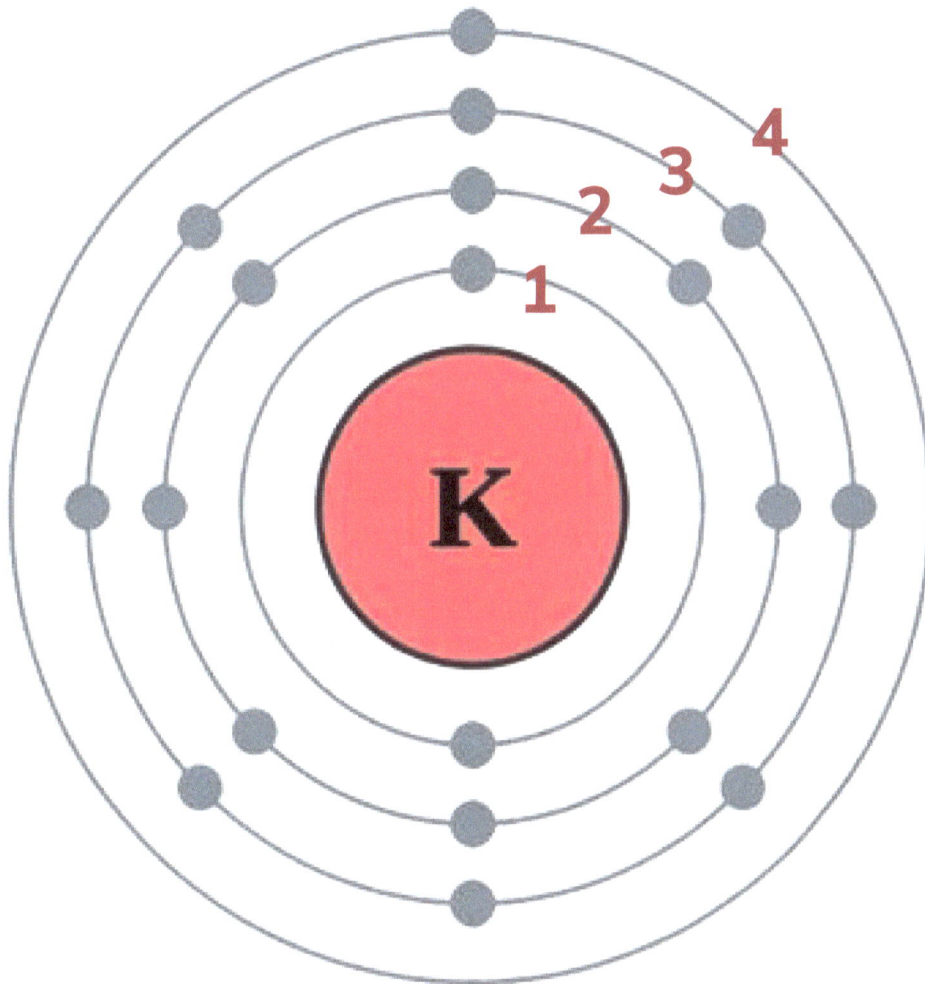

Hi I'm Mr. Eddie. I teach period 6.

All of the elements in my class have 6 outer energy shells.

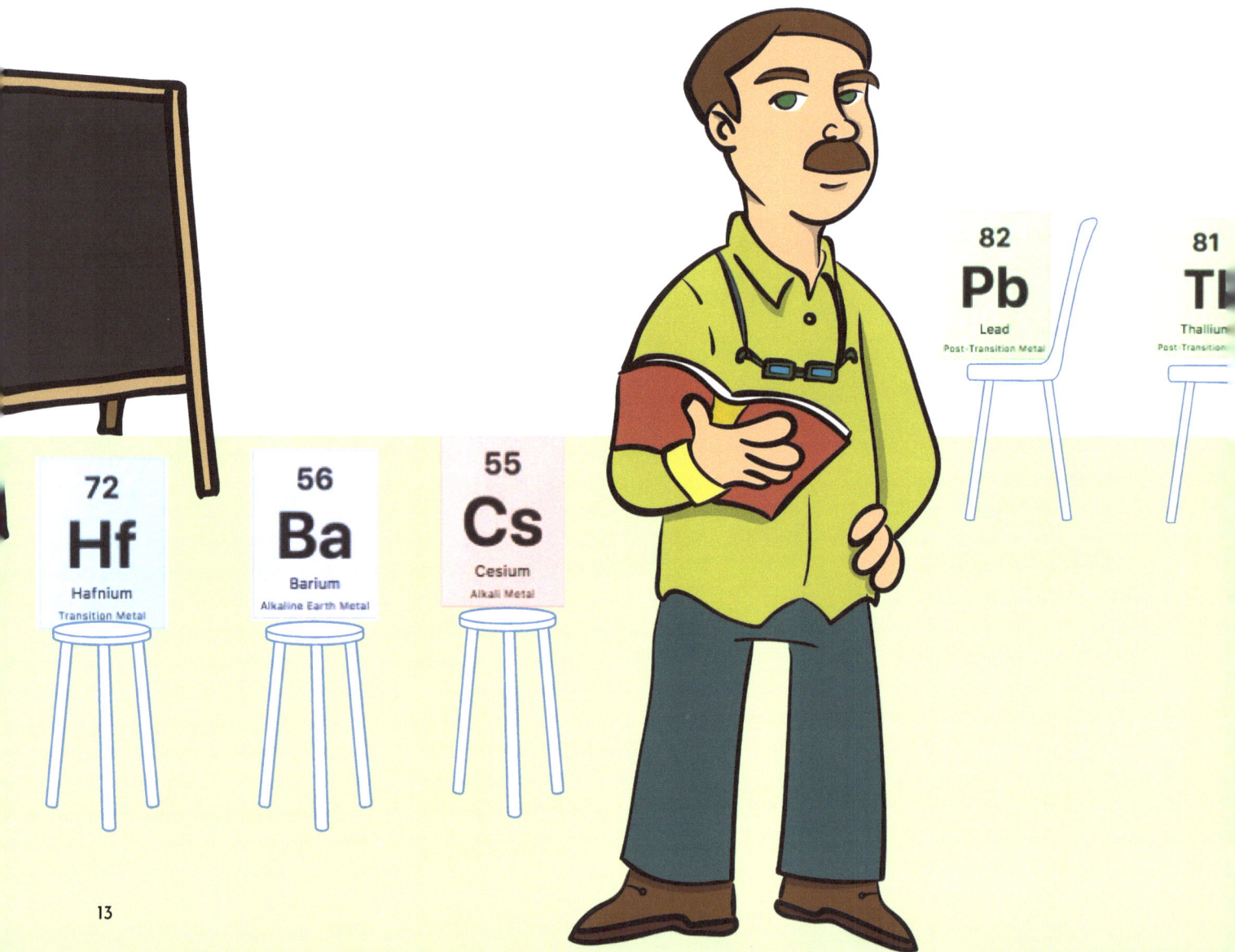

72
Hf
Hafnium
Transition Metal

56
Ba
Barium
Alkaline Earth Metal

55
Cs
Cesium
Alkali Metal

82
Pb
Lead
Post-Transition Metal

81
Tl
Thallium
Post-Transition

Count the outer shells!

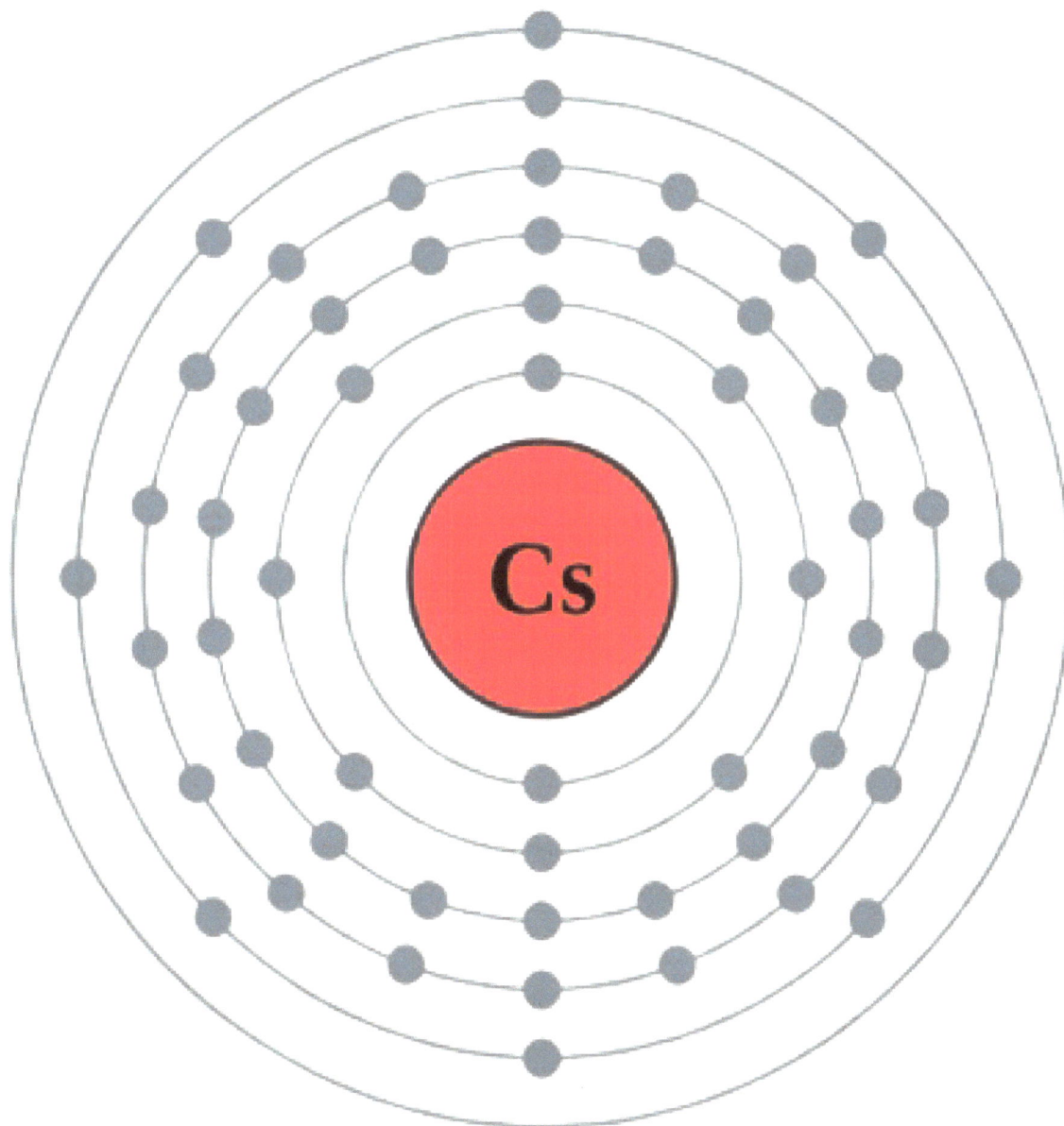

Let's check out the cafeteria!

81
Tl
Thallium
Post-Transition Metal

8
O
Oxygen
Nonmetal

4
Be
Beryllium
Alkaline Earth Metal

7
N
Nitrogen
Nonmetal

73
Ta
Tantalum
Transition Metal

20
Ca
Calcium
Alkaline Earth Metal

Elements sit with other elements that function similarly to them. For example, Nitrogen, Arsenic and Phosphorus all sit at table 5 because they're all transition metals.

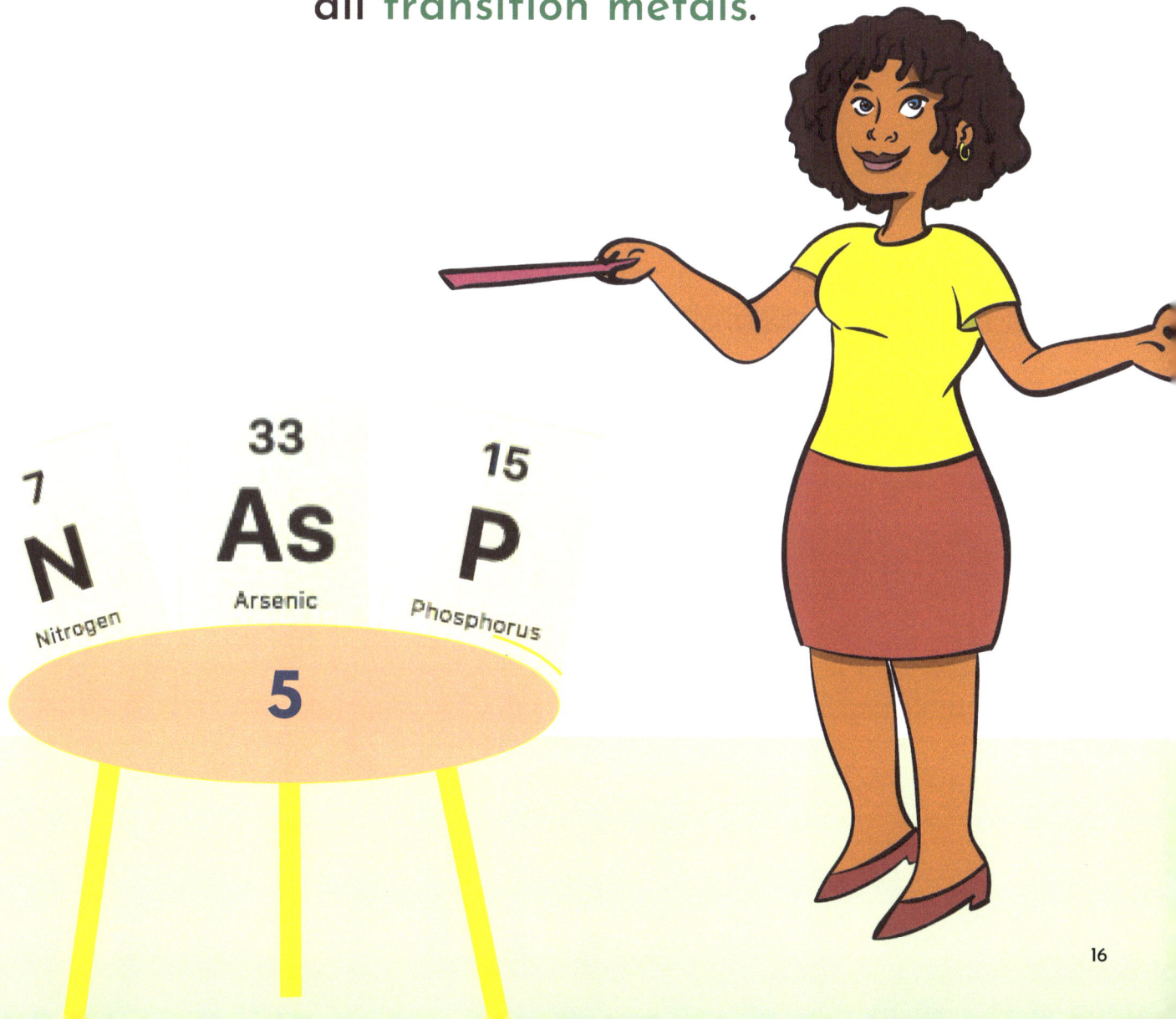

7

N

Nitrogen

33

As

Arsenic

15

P

Phosphorus

5

At table 8 you can find Xenon, Argon, Neon, Helium and Radon. They sit in a group because they are all noble gases.

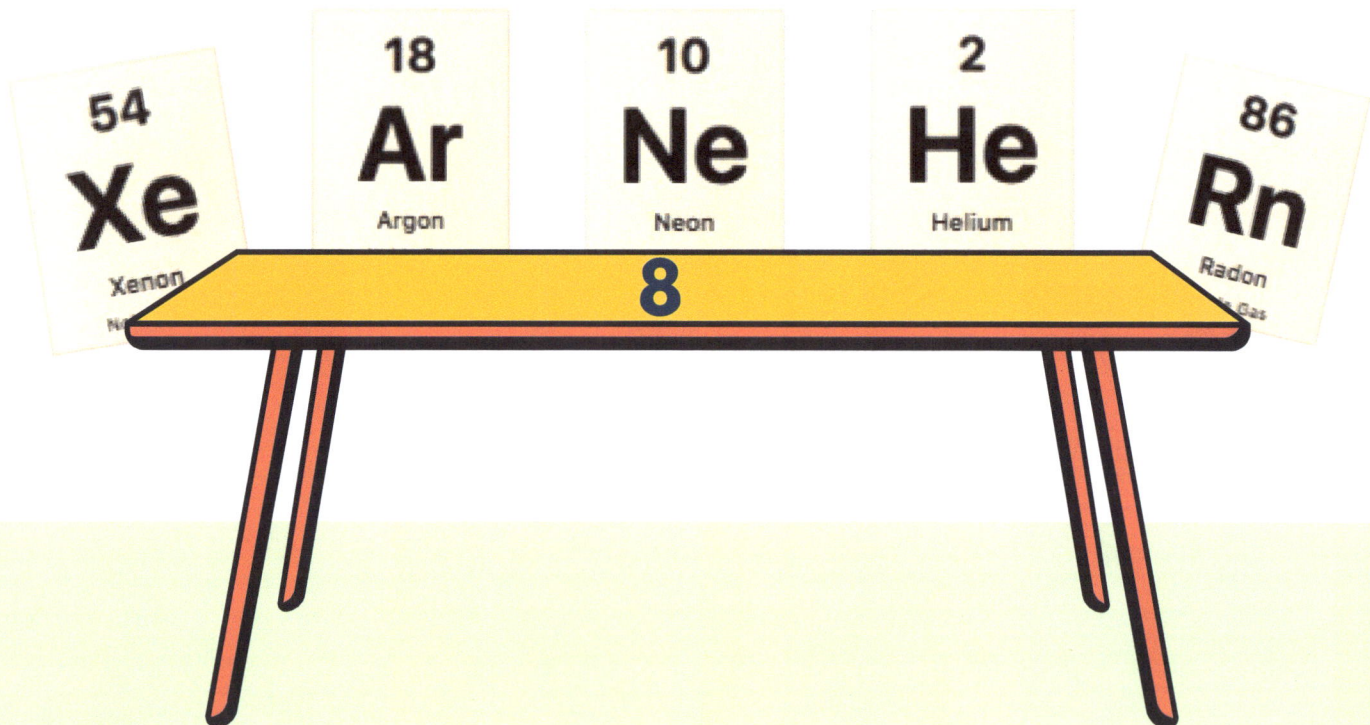

Oh! I forgot to tell you, today is picture day.

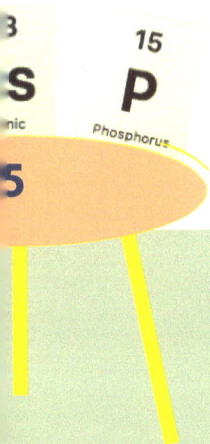

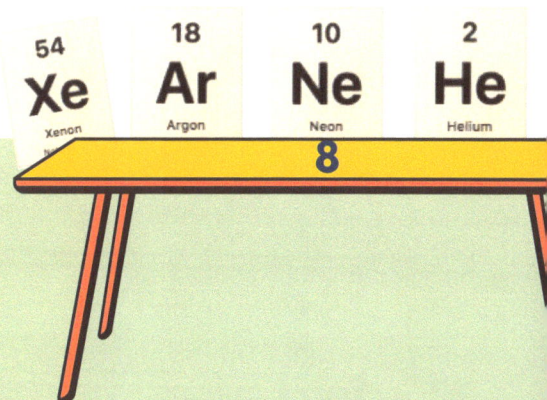

Everyone say Periodic Tableeeee!

By taking a closer look at our school picture, you can see the different periods and groups. Periods are identified by looking at the chart's rows. Numbered 1 - 7.

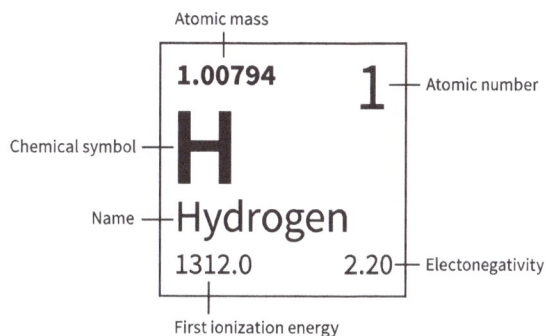

Atomic mass

1.00794

1 — Atomic number

Chemical symbol — H

Name — Hydrogen

1312.0   2.20 — Electonegativity

First ionization energy

Alkali metals

Transition metals

Metalloids

Alkaline metals

Lanthanoids

Nonmetals

Other metals

Actinoids

Halogens

Noble gases

The Periodic Table of Elements

Periods are identified by looking at the chart's rows numbered 1 - 7

	1 IA 11A	1 IIA 2A	3 IIIB 3B	4 IVB 4B	5 VB 5B	6 VIB 6B	7 VIIB 7B	8 VIII 8	9 VIII 8	10 VIII 8	11 IB 1B	12 IIB 2B	13 IIIA 3A	14 IVA 4A	15 VA 5A	16 VIA 6A	17 VIIA 7A	
1	1.00794 **1** H Hydrogen 1312.0 2.20																	4.00 **H** Hel 2372
2	6.941 **3** Li Lithium 520.2 0.98	9.012182 **4** Be Beryllium 899.5 1.57											10.811 **5** B Boron 800.6 2.04	12.0107 **6** C Carbon 1086.5 2.55	14.0067 **7** N Nitrogen 1402.3 3.04	15.9994 **8** O Oxygen 1313.9 3.44	18.998403 **9** F Fluorine 1681 3.98	20.1 **N** Ne 2080
3	22.98976 **11** Na Sodium 495.8 0.93	24.3050 **12** Mg Magnesium 737.7 1.31											26.98153 **13** Al Aluminium 577.5 1.61	28.0855 **14** Si Silicon 786.5 1.90	30.97696 **15** P Phosphorus 1011.8 2.19	32.065 **16** S Sulfur 999.6 2.58	35.453 **17** Cl Chlorine 1251.2 3.16	39.9 **A** A 1520
4	39.0983 **19** K Potassium 418.8 0.82	40.078 **20** Ca Calcium 589.8 1.00	44.95591 **21** Sc Scandium 633.1 1.36	47.867 **22** Ti Titanium 658.8 1.54	50.9415 **23** V Vanadium 650.9 1.63	51.9962 **24** Cr Chromium 652.9 1.66	54.93804 **25** Mn Manganese 717.3 1.55	55.845 **26** Fe Iron 762.5 1.83	58.93319 **27** Co Cobalt 700.4 1.88	58.6934 **28** Ni Nickel 737.1 1.91	63.546 **29** Cu Copper 745.5 1.90	65.38 **30** Zn Zinc 906.4 1.65	69.723 **31** Ga Gallium 578.8 1.81	72.64 **32** Ge Germanium 762 2.01	74.92160 **33** As Arsenic 947 2.18	78.96 **34** Se Selenium 941 2.55	79.904 **35** Br Bromine 1139.9 2.96	83.7 **K** Kry 1350
5	85.4678 **37** Rb Rubidium 403 0.82	87.62 **38** Sr Strontium 549.5 0.95	88.90585 **39** Y Yttrium 600 1.22	91.224 **40** Zr Zirconium 640.1 1.33	92.90638 **41** Nb Niobium 652.1 1.60	95.96 **42** Mo Molybdenum 684.3 2.16	(98) **43** Tc Technetium 702 1.90	101.07 **44** Ru Ruthenium 710.2 2.2	102.9055 **45** Rh Rhodium 719.7 2.28	106.42 **46** Pd Palladium 804.4 2.20	107.8682 **47** Ag Silver 731 1.93	112.441 **48** Cd Cadmium 867.8 1.69	114.818 **49** In Indium 558.3 1.78	118.710 **50** Sn Tin 708.6 1.96	121.760 **51** Sb Antimony 834 2.05	127.60 **52** Te Tellurium 869.3 2.10	126.9044 **53** I Iodine 1008.4 2.66	131.2 **X** Xen 1170
6	132.9054 **55** Cs Caesium 375.7 0.79	137.327 **56** Ba Barium 502.9 0.89	174.9668 **71** Lu Lutetium 523.5 1.27	178.49 **72** Hf Hafnium 658.5 1.30	180.9478 **73** Ta Tantalum 761 1.50	183.84 **74** W Tungsten 770 2.36	186.207 **75** Re Rhenium 760 1.90	190.23 **76** Os Osmium 840.0 2.2	192.217 **77** Ir Iridium 880 2.20	195.084 **78** Pt Platinum 870 2.28	196.9665 **79** Au Gold 890.1 2.54	200.59 **80** Hg Mercury 1007.1 2.00	204.3833 **81** Tl Thallium 589.4 1.62	207.2 **82** Pb Lead 715.6 2.33	208.9804 **83** Bi Bismuth 703 2.02	(210) **84** Po Polonium 890 2.00	(210) **85** At Astatine 890 2.20	(220) **R** Rac 1037
7	(223) **87** Fr Francium 380.0 0.7	(226) **88** Ra Radium 509.3 0.90	(262) **103** Lr Lawrencium 470	(261) **104** Rf Rutherfordium 580	(262) **105** Db Dubnium	(266) **106** Sg Seaborgium	(264) **107** Bh Bohrium	(277) **108** Hs Hassium	(268) **109** Mt Meitnerium	(271) **110** Ds Darmstadium	(272) **111** Rg Roentgenium	(285) **112** Cn Copernicium	(284) **113** Uut Ununtrium	(289) **114** Fl Flerovium	(288) **115** Uup Ununpentium	(292) **116** Lv Livermorium	(294) **117** Uus Ununseptium	(294) **U** Unu

Atomic mass → 1.00794 **1** ← Atomic number
Chemical symbol → **H**
Name → Hydrogen
1312.0 2.20 ← Electronegativity
First Ionization energy ↑

138.9054 **57** La Lanthanum 538.1 1.10	140.116 **58** Ce Cerium 534.4 1.12	140.9076 **59** Pr Praseodymium 527 1.13	144.242 **60** Nd Neodymium 533.1 1.14	(145) **61** Pm Promethium 540	150.36 **62** Sm Samarium 544.5 1.17	151.964 **63** Eu Europium 547.1	157.25 **64** Gd Gadolinium 593.4 1.20	158.9253 **65** Tb Terbium 565.8	162.500 **66** Dy Dysprosium 573 1.22	164.9303 **67** Ho Holmium 581 1.23	167.259 **68** Er Erbium 589.3 1.24	168.9342 **69** Tm Thulium 596.7 1.25	173.0 **Y** Ytte 603.4
(227) **89** Ac Actinium 499 1.10	232.0380 **90** Th Thorium 587 1.30	231.0358 **91** Pa Protactinium 508 1.50	238.0289 **92** U Uranium 597.6 1.38	(237) **93** Np Neptunium 604.5 1.36	(244) **94** Pu Plutonium 584.7 1.28	(243) **95** Am Americium 576 1.30	(247) **96** Cm Curium 581 1.30	(247) **97** Bk Berkelium 601 1.30	(251) **98** Cf Californium 608 1.30	(252) **99** Es Einsteinium 619 1.30	(257) **100** Fm Fermium 627 1.30	(258) **101** Md Mendelevium 635 1.30	(259) **N** Nob 642

| Alkali metals | Alkaline metals | Other metals | Transition metals | Lanthanoids | Actinoids | Metalloids | Nonmetals | Halogens | |

FUN FACT

How are the colors on the periodic table chosen?

There is no standard set of colors used to identify element groups or other properties. Colors are selected based on how well the text shows up against them, but mostly it's a matter of personal preference. You can find periodic tables in a variety of color schemes.

Groups are determined by the charts columns numbered **1-18.** As we saw in the cafeteria, elements that share the same group, sit at the same table. When elements share the same group that means they have the same number of electrons in their most outer shell and act similarly to each other.

Groups

1 ... 18

Periods

These are also a part of period 7

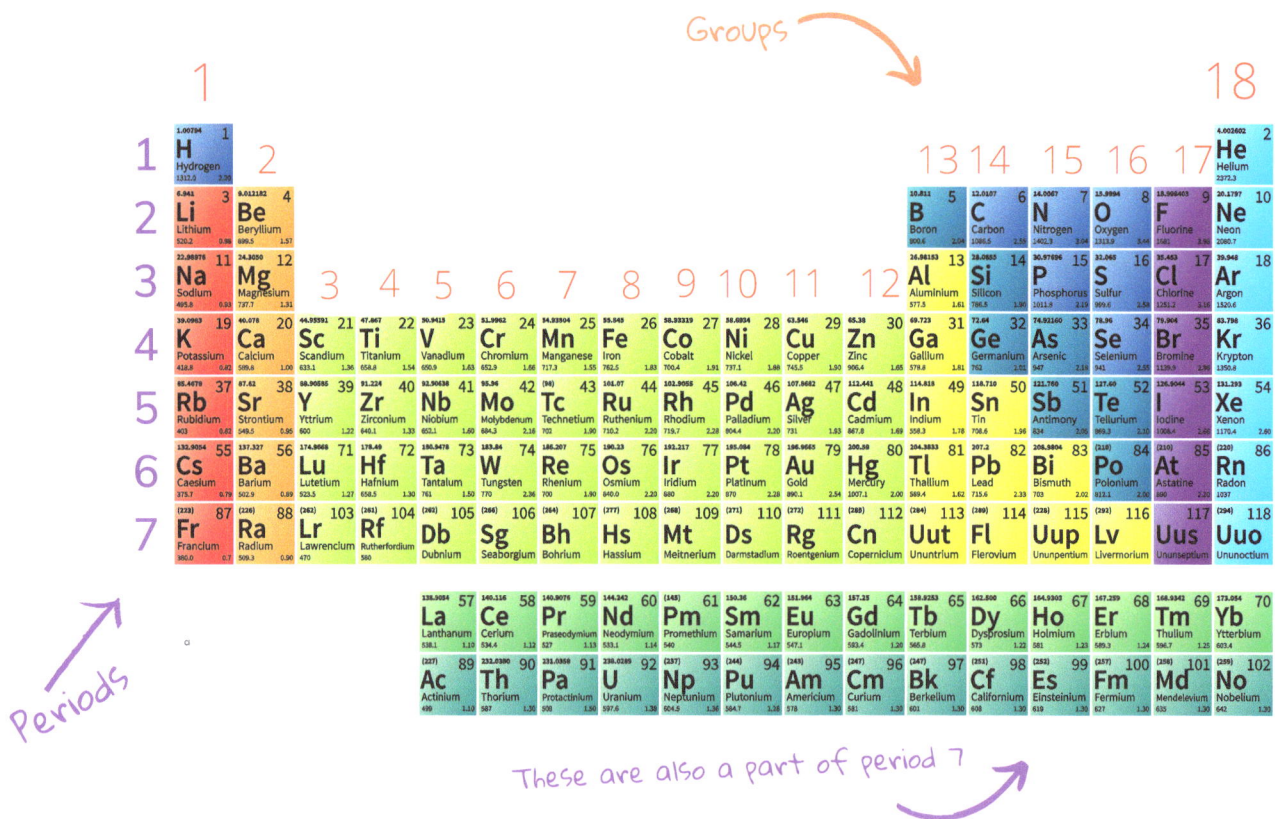

aluminum

Al

13

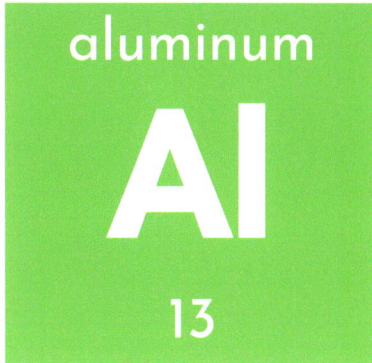

Let's take a look at one element on the chart: Al

The element with the symbol Al is Aluminum. By looking at the chart, we can tell Aluminum has 3 shells and 3 electrons on it's most outer shell. Aluminum belongs to group 3 meaning it's a part of the Scandium Family.

The elements in group 3 are considered poor metals. These elements are softer and have a low melting point.

Some other elements in this group are boron, gallium and indium.

FUN FACT

A <u>group</u> <u>is a vertical column of the periodic table.</u>

There are a total of 18 groups. There are two different numbering systems that are commonly used. The traditional system used in the United States and a different system in Europe. To eliminate confusion the International Union of Pure and Applied Chemistry (IUPAC) decided that the official system for numbering groups would be a simple 1 through 18 from left to right. Though many periodic tables show both systems simultaneously.

It is important to know how to read a periodic table when studying chemistry. The periodic table is basically the guide to understanding the complex environment around us.

I hope you learned something new while visiting us! Thanks for coming and see you soon!

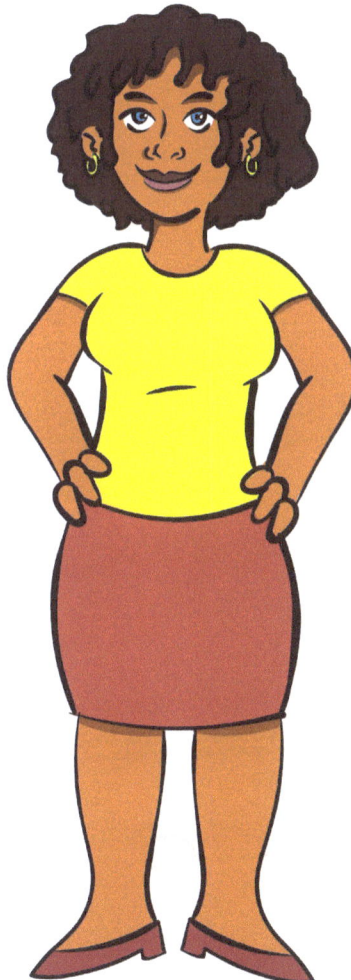

Groups

Group 1: alkali metals, or lithium family.

Group 2: alkaline earth metals, or beryllium family.

Group 3: scandium family plus rare earth metals.

Group 4: the titanium family.

Group 5: the vanadium family.

Group 6: the chromium family.

Group 7: the manganese family.

Group 8: the iron family.

Group 9: cobalt family.

Group 10: nickel family.

Group 11: copper family.

Group 12: zinc family.

Group 13: boron family.

Group 14: carbon family.

Group 15: nitrogen family.

Group 16: oxygen family.

Group 17: halogens or fluoride family.

Group 18: noble gases or helium family or neon family.

* Transition Metals are in Green

Glossary

Chemical Element: Any substance that cannot be decomposed into simpler substances by ordinary chemical processes. Elements are the fundamental materials of which all matter is composed.

Noble Gas: Any of the seven chemical elements that make up Group 18 of the periodic table.

Poor Metal: Any of various metallic elements generally with higher electronegativity, lower melting and boiling points and greater softness than the transition metals.

Principal Energy Level: Refers to the shell in which the electron is located relative to the atom's nucleus. The first element in a period of the periodic table introduces a new principal energy level.

Transition Metal: Any of various metallic elements (such as chromium, iron, and nickel) that have valence electrons in two shells instead of only one.

Valence Electron: An outer shell electron that is associated with an atom, and it can help form a chemical bond if the outer shell is not closed.

About The Author

Born and raised in the Bronx, NY, Jada Camille is a 16-year-old writer and author. She uses her passion for storytelling to help simplify concepts to engage her readers. Jada fell in love with writing and storytelling in the 3rd grade after a literature project assigned by her teacher revealed her writing talent; and yes, she got an "A." Since then, Jada has gone on to have her stories featured in education and media outlets. When she's not writing, Jada spends time listening to music, hanging with her family, and getting her feet wet in modeling & beauty ventures. She hopes that her writing will inspire other young people to believe in their power to do big things no matter who and where they are, as she considers herself to be just a normal teenager who happens to be a writer.

You can sign up for her mailing list and receive updates on her next manuscript at manager@leptalent.com.

www.ingramcontent.com/pod-product-compliance
Lightning Source LLC
Chambersburg PA
CBHW052044190326
41520CB00002BA/188